THE PRACTICE OF BUSINESS STATISTICS

COMPANION CHAPTER 15
TWO-WAY ANALYSIS OF VARIANCE

David S. Moore
Purdue University

George P. McCabe
Purdue University

William M. Duckworth
Iowa State University

Stanley L. Sclove
University of Illinois

W. H. Freeman and Company
New York

Senior Acquisitions Editor:	**Patrick Farace**
Senior Developmental Editor:	**Terri Ward**
Associate Editor:	**Danielle Swearengin**
Media Editor:	**Brian Donnellan**
Marketing Manager:	**Jeffrey Rucker**
Head of Strategic Market Development:	**Clancy Marshall**
Project Editor:	**Mary Louise Byrd**
Cover and Text Design:	**Vicki Tomaselli**
Production Coordinator:	**Paul W. Rohloff**
Composition:	**Publication Services**
Manufacturing:	**RR Donnelley & Sons Company**

Cataloguing-in-Publication Data available from the Library of Congress

Library of Congress Control Number: 2002108463

Printed in the United States of America

First Printing 2002

TO THE INSTRUCTOR

NOW *YOU* HAVE THE CHOICE!

This is **Companion Chapter 15** to *The Practice of Business Statistics (PBS)*. Please note that this chapter, along with any other Companion Chapters, can be bundled with the *PBS* Core book, which contains Chapters 1–11.

CORE BOOK

Chapter 1	Examining Distributions
Chapter 2	Examining Relationships
Chapter 3	Producing Data
Chapter 4	Probability and Sampling Distributions
Chapter 5	Probability Theory
Chapter 6	Introduction to Inference
Chapter 7	Inference for Distributions
Chapter 8	Inference for Proportions
Chapter 9	Inference for Two-Way Tables
Chapter 10	Inference for Regression
Chapter 11	Multiple Regression

These other **Companion Chapters**, *in any combinations you wish,* are available for you to package with the *PBS* Core book.

TAKE YOUR PICK

Chapter 12	Statistics for Quality: Control and Capability
Chapter 13	Time Series Forecasting
Chapter 14	One-Way Analysis of Variance
Chapter 15	Two-Way Analysis of Variance
Chapter 16	Nonparametric Tests
Chapter 17	Logistic Regression
Chapter 18	Bootstrap Methods and Permutation Tests

TWO-WAY ANALYSIS OF VARIANCE

Prelude: One-way ANOVA or two-way ANOVA?

A researcher for a company that markets dietary supplements wants to study the effect of calcium intake on blood pressure. She plans to perform an experiment with three groups of rats, each group fed identical diets except for the amount of calcium. Call the levels of calcium low, medium, and high. Previous rat studies allow a rough estimate of the standard deviation of the response variable, which is blood pressure. Calculations of power then show that 15 rats per group should allow detection of differences of the size the researcher considers important. This experiment fits the model for one-way ANOVA.

The researcher also plans a similar study on the effect of magnesium. Unlike the calcium study, she expects that magnesium will have little effect on the rats' blood pressure. Should she conduct a separate experiment with a second one-way ANOVA? Rather than two experiments, each with 45 rats, a statistician suggests that the experiments be combined. Here's the new design to investigate the effects of both calcium and magnesium: Use three levels (low, medium, high) of each supplement. Pair each calcium level with each magnesium level, giving a total of 3×3, or 9, diets. Randomly assign 9 rats to each diet, a total of 81 rats. The analysis for this experiment is a *two-way ANOVA*. In this chapter we will learn some details of two-way ANOVA and its advantages in situations such as this.

CHAPTER 15

Two-Way Analysis of Variance

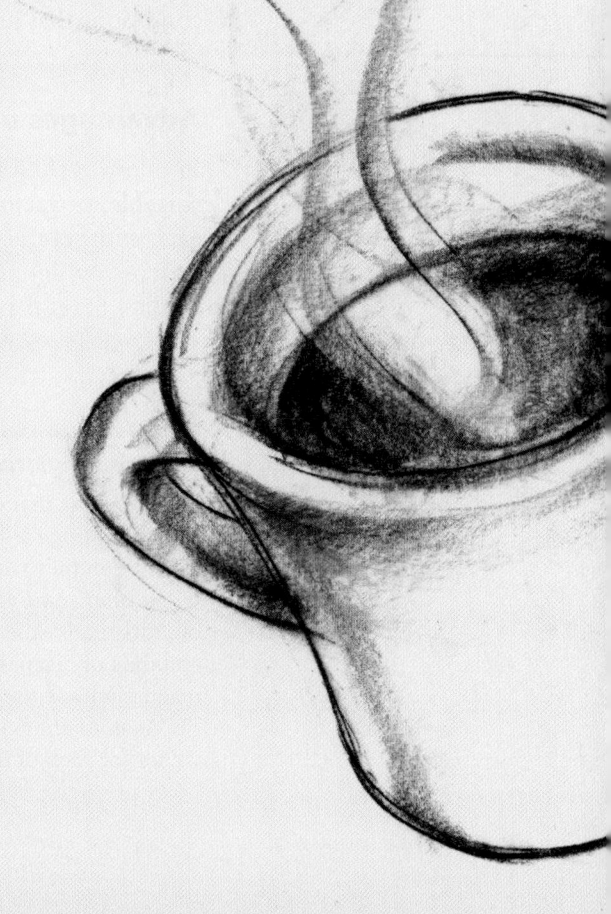

Introduction

The two-sample t procedures of Chapter 7 compare the means of two populations. We generalized these in Chapter 14 to one-way analysis of variance (ANOVA) for comparing the means of several populations.

two-way ANOVA

Two-way ANOVA compares the means of populations that are classified in two ways or the mean responses in two-factor experiments. Many of the key concepts are similar to those of one-way ANOVA, but the presence of more than one factor also introduces some new ideas. We once more assume that the data are approximately Normal and that groups may have different means but have the same standard deviation; we again pool to estimate the variance; and we again use F statistics for significance tests. The major difference between one-way and two-way ANOVA is in the FIT part of the model. We will carefully study this term, and we will find much that is both new and useful.

higher-way ANOVA

We may, of course, have more than two factors. The statistical analysis is then called **higher-way ANOVA.** Although some details grow more complex, the most important ideas are already present in the two-way setting.

15.1 The Two-Way ANOVA Model

We begin with a discussion of the advantages of the two-way ANOVA design and illustrate these with some examples. Then we discuss the model and the conditions that justify two-way ANOVA.

Advantages of two-way ANOVA

factor

In one-way ANOVA, we classify populations according to one categorical variable, or **factor.** In the two-way ANOVA model, there are two factors, each with several levels. When we are interested in the effects of two factors, a two-way design offers great advantages over two or more single-factor studies. Several examples will illustrate these advantages.

EXAMPLE 15.1 **Dietary supplements and blood pressure**

The company that conducted the study described in the Prelude was interested in demonstrating the beneficial effect of their calcium supplement on blood pressure. In addition they wanted to explore the possibility that a magnesium supplement might offer some added benefit. In the two-factor experiment the researcher settled on, rats receive diets prepared with varying amounts of calcium and varying amounts of magnesium, but otherwise identical. Amount of calcium and amount of magnesium are the two factors in this experiment.

As is common in such experiments, the researcher set high, medium, and low values for each of the two minerals. So there are three levels for each of the factors and a total of nine diets, or treatments. The following table summarizes the factors and levels:

	Calcium		
Magnesium	L	M	H
L	1	2	3
M	4	5	6
H	7	8	9

For example, Diet 2 contains magnesium at its low level combined with the normal (medium) level of calcium. Each diet is fed to 9 rats, giving a total of 81 rats used in this experiment. The response variable is the blood pressure of a rat after some time on the diet.[1]

The researcher in this example is primarily interested in the effect of calcium on blood pressure. She might conduct a single-factor experiment with magnesium set at its normal (medium) value and three levels of calcium, using one-way ANOVA to analyze the responses. If she then decides to study the effect of magnesium, she might perform another experiment with calcium set at its normal value and three levels of magnesium. Suppose that the budget allows for only 80 to 90 rats in all. For the first experiment, the researcher might assign 15 rats to each of the three calcium diets. Similarly, she could assign 15 rats to each of the three magnesium diets in the second experiment. This would require a total of 90 rats, 9 more than the two-factor design in Example 15.1. Moreover, some combinations, such as high calcium with high magnesium, appear in neither experiment.

In the two-way experiment, all 9 calcium-magnesium combinations appear. Moreover, 27 rats are assigned to each of the three calcium levels and 27 rats to each of the three magnesium levels. That is, the two-way experiment provides more information about both minerals than the two one-way studies with 15 rats per mineral level. We can express this fact using standard deviations of the mean responses. If σ is the standard deviation of blood pressure for rats fed the same diet, then the standard deviation of a sample mean for n rats is $\sigma/\sqrt{n}$. For the two-way experiment this is $\sigma/\sqrt{27} = 0.19\sigma$, whereas for the two one-way experiments we have $\sigma/\sqrt{15} = 0.26\sigma$. The sample mean responses for any one level of either calcium or magnesium are less variable in the two-way design.

When the actual experiment was analyzed using two-way ANOVA, no effect of magnesium on blood pressure was found. On the other hand, the data clearly demonstrated that the high-calcium diet had the beneficial effect of lowering the average blood pressure. Means for the three calcium levels were reported with standard errors based on 27 rats per group.

The conclusions about the effect of calcium on blood pressure are also more general in the two-way experiment. In the one-way experiments, calcium is varied only for the normal level of magnesium. All levels of calcium are combined with all levels of magnesium in the two-way design, so that conclusions about calcium are not confined to one level of magnesium. The same increased generality also applies to conclusions about magnesium. So the conclusions are both more precise and more general, yet the two-way design requires fewer rats. Experiments with several factors are an efficient use of resources.

EXAMPLE 15.2 **Designing a student alcohol abuse survey**

ANOVA methods apply to sample surveys as well as to experiments. A group of students and administrators concerned about abuse of alcohol is planning an education and awareness program for the student body at a large university. To aid in the design of their program they decide to gather information on the attitudes of undergraduate students toward alcohol.

Because the attitudes of freshmen, sophomores, juniors, and seniors may be quite different, a stratified sample is appropriate. The study planners decide to use year in school as a factor with four levels and to take separate random samples from each class of students.

It is also possible that men and women have different attitudes toward alcohol. The team therefore decides to add gender as a second factor. The final sampling design has two factors: year with four levels, and gender with two levels. An SRS of size 50 will be drawn from each of the eight populations. The eight groups for this study are shown in the following table:

	Year			
Gender	Fr.	So.	Jr.	Sr.
Male	1	2	3	4
Female	5	6	7	8

Example 15.2 illustrates another advantage of two-way designs analyzed by two-way ANOVA. In the one-way design with year as the only factor, all four random samples would include both men and women. If a gender difference exists, the one-way ANOVA would assign this variation to the RESIDUAL (within groups) part of the model. In the two-way ANOVA, gender is included as a factor, and therefore this variation is included in the FIT part of the model. Whenever we can move variation from RESIDUAL to FIT, we reduce the σ (within-group variation) of our data and increase the power of our tests.

EXAMPLE 15.3 **Durability of fabrics**

A researcher for a company that sells textiles is interested in how four different colors of dye affect the durability of fabrics. Because the effects of the dyes may be different for different types of cloth, he applies each dye to five different kinds of cloth. The two factors in this experiment are dyes with four levels and cloth types with five levels. Six fabric specimens are dyed for each of the 20 dye-cloth combinations, and all 120 specimens are tested for durability.

The researcher in Example 15.3 could have used only one type of cloth and compared the four dyes using a one-way analysis. However, based on knowledge of the chemistry of textiles and dyes, he suspected that some of the dyes might have a chemical reaction with some types of cloth. Such a reaction could have a strong negative effect on the durability of the cloth. He wanted his results to be useful to manufacturers who would use the dyes on a variety of cloths. He chose several cloth types for the study to represent the different kinds that would be used with these dyes.

interactions

main effects

If one or more specific dye-cloth combinations produce exceptionally bad or exceptionally good durability measures, the experiment should discover this combined effect. Effects of the dyes that differ for different types of cloth are represented in the FIT part of a two-way model as **interactions.** In contrast, the average values for dyes and cloths are represented as **main effects.** The two-way model represents FIT as the sum of a main effect for each of the two factors *and* an interaction. One-way designs that vary a single factor and hold other factors fixed cannot discover interactions. We will discuss interactions more fully in a later section.

These examples illustrate several reasons why two-way designs are preferable to one-way designs.

ADVANTAGES OF TWO-WAY ANOVA

1. It is more efficient to study two factors simultaneously rather than separately.
2. We can reduce the residual variation in a model by including a second factor thought to influence the response.
3. We can investigate interactions between factors.

These considerations also apply to study designs with more than two factors. We will be content to explore only the two-way case. The choice of the design for data production (sample or experiment) is fundamental to any statistical study. Factors and levels must be carefully selected by an individual or team who understands both the statistical models and the issues that the study will address.

The two-way ANOVA model

When discussing two-way models in general, we will use the labels A and B for the two factors. For particular examples and when using statistical software, it is better to use names for these categorical variables that suggest their meaning. Thus, in Example 15.1 we would say that the factors are calcium and magnesium. The numbers of levels of the factors are often used to describe the model. Again using Example 15.1 as an illustration, we would call this a 3×3 ANOVA. Similarly, Example 15.2 illustrates a 2×4 ANOVA. In general, factor A will have I levels and factor B will have J levels. Therefore, we call the general two-way problem an $I \times J$ ANOVA.

In a two-way design every level of A appears in combination with every level of B, so that $I \times J$ groups are compared. The sample size for level i of factor A and level j of factor B is n_{ij}.[2]

The total number of observations is

$$N = \sum n_{ij}$$

A single response variable is measured for each observation.

ASSUMPTIONS FOR TWO-WAY ANOVA

We have independent SRSs of size n_{ij} from each of $I \times J$ Normal populations. The population means μ_{ij} may differ, but all populations have the same standard deviation σ. The μ_{ij} and σ are unknown parameters.

Let x_{ijk} represent the kth observation from the population having factor A at level i and factor B at level j. The statistical model is

$$x_{ijk} = \mu_{ij} + \epsilon_{ijk}$$

for $i = 1, \ldots, I$ and $j = 1, \ldots, J$ and $k = 1, \ldots, n_{ij}$. The deviations ϵ_{ijk} are from an $N(0, \sigma)$ distribution.

The FIT part of the model is the means μ_{ij}, and the RESIDUAL part is the deviations ϵ_{ijk} of the individual observations from their group means. To estimate a population mean μ_{ij} we use the sample mean of the observations in the sample from this group:

$$\overline{x}_{ij} = \frac{1}{n_{ij}} \sum_{k} x_{ijk}$$

The k below the $\sum$ means that we sum the n_{ij} observations that belong to the (i, j)th sample.

The RESIDUAL part of the model contains the unknown σ. We calculate the sample variances for each SRS and pool these to estimate σ^2:

$$s_p^2 = \frac{\sum (n_{ij} - 1) s_{ij}^2}{\sum (n_{ij} - 1)}$$

pooled standard error

Just as in one-way ANOVA, the numerator in this fraction is SSE and the denominator is DFE. Also as in the one-way analysis, DFE is the total number of observations minus the number of groups. That is, DFE $= N - IJ$. The estimator of σ is s_p, the **pooled standard error.**

15.1 **Compare employee training programs.** A company wants to compare three different training programs for its new employees. Each of these programs takes 8 hours to complete. The training can be given for 8 hours on one day or for 4 hours on two consecutive days. The next 120 employees that the company hires will be the subjects for this study. After the training is completed, the employees are asked to evaluate the effectiveness of the program on a 7-point scale. Identify the response variable and both factors, and state the number of levels for each factor (I and J) and the total number of observations (N).

15.2 **Are some colors more attractive to customers?** A marketing experiment compares four different types of packaging for computer disks. Each type of packaging can be presented in three different colors. Each combination of package type with a particular color is shown to 40 different potential customers, who rate the attractiveness of the product on a 1 to 10 scale. Describe the two factors, and give the number of levels of each and the total number of observations. What is the response variable?

15.3 **Which hand lotion should you market?** Five different formulations for your hand lotion product have been produced by your research and marketing group, and you want to decide which of these, if any, to market. The lotions can be made with three different fragrances. Samples of each formulation-by-fragrance lotion are sent to 120 randomly selected customers who use your regular product. You ask each customer to compare the new lotion with the regular product by rating it on a 7-point scale. The middle point of the scale corresponds to no preference, while higher values indicate that the new product is preferred and lower values indicate that the regular product is better. What is the response variable? Give the factors with numbers of levels and the total sample size.

Main effects and interactions

The first important distinction between one-way and two-way ANOVA is that in two-way ANOVA we break down the FIT part of the model (the population means μ_{ij}) in a way that reflects the presence of two factors.

Because we have independent samples from each of $I \times J$ groups, we can think of the problem initially as a one-way ANOVA with IJ groups. Each population mean μ_{ij} is estimated by the corresponding sample mean $\bar{x}_{ij}$. We can calculate sums of squares and degrees of freedom as in one-way ANOVA. In accordance with the conventions used by many computer software packages, we use the term *model* when discussing the sums of squares and degrees of freedom calculated as in one-way ANOVA with IJ groups. Thus, SSM is a model sum of squares constructed from deviations of the form $\bar{x}_{ij} - \bar{x}$, where $\bar{x}$ is the average of all of the observations and $\bar{x}_{ij}$ is the mean of the (i, j)th group. Similarly, DFM is simply $IJ - 1$.

In two-way ANOVA, the terms SSM and DFM are broken down into terms corresponding to a main effect for A, a main effect for B, and an AB interaction. Each of SSM and DFM is then a sum of terms:

$$\text{SSM} = \text{SSA} + \text{SSB} + \text{SSAB}$$

and

$$\text{DFM} = \text{DFA} + \text{DFB} + \text{DFAB}$$

The term SSA represents variation among the means for the different levels of the factor A. Because there are I such means, DFA $= I - 1$ degrees of freedom. Similarly, SSB represents variation among the means for the different levels of the factor B, with DFB $= J - 1$.

Interactions are a bit more involved. We can see that SSAB, which is SSM $-$ SSA $-$ SSB, represents the variation in the model that is not accounted for by the main effects. By subtraction we see that its degrees of freedom are

$$\text{DFAB} = (IJ - 1) - (I - 1) - (J - 1) = (I - 1)(J - 1)$$

There are many kinds of interactions. The easiest way to study them is through examples.

EXAMPLE 15.4 Workers' earnings

Each March, the Bureau of Labor Statistics collects detailed data on incomes. Here, from the March 2001 survey, are the mean earned incomes of all men and women in two age groups in the year 2000.[3] (The data include people who earned nothing.)

Age	Women	Men	Mean
15–24	\$10,744	\$13,178	\$11,961
25–44	\$27,248	\$44,224	\$35,736
Mean	\$18,996	\$28,701	\$23,848.5

The table includes averages of the means in the rows and columns. For example, the first entry in the far right margin is the average of the salaries for women aged 15 to 24 and men aged 15 to 24:

$$\frac{10{,}744 + 13{,}178}{2} = 11{,}961$$

Similarly, the average of women's salaries in the two age groups is

$$\frac{10{,}744 + 27{,}248}{2} = 18{,}996$$

marginal means

These averages are called **marginal means** because of their location at the margins of the table.

main effects

It is clear from the marginal means that men earn more than women on the average and that people in the older age group earn more than those in the younger group. These are **main effects** for the two factors. We can describe the main effects by the differences between the marginal means. Men earn an average of \$9705 more than women, and people aged 25 to 44 earn an average of \$23,775 more than people aged 15 to 24.

interaction

What about the interaction between gender and age? An **interaction** is present if the main effects provide an incomplete description of the data. That is, if the male-female earnings gap is different in the two age groups, then gender and age interact. In fact, the gap is much larger in the older age group:

	15 to 24 years	25 to 44 years
Male-female difference	\$2,434	\$16,976

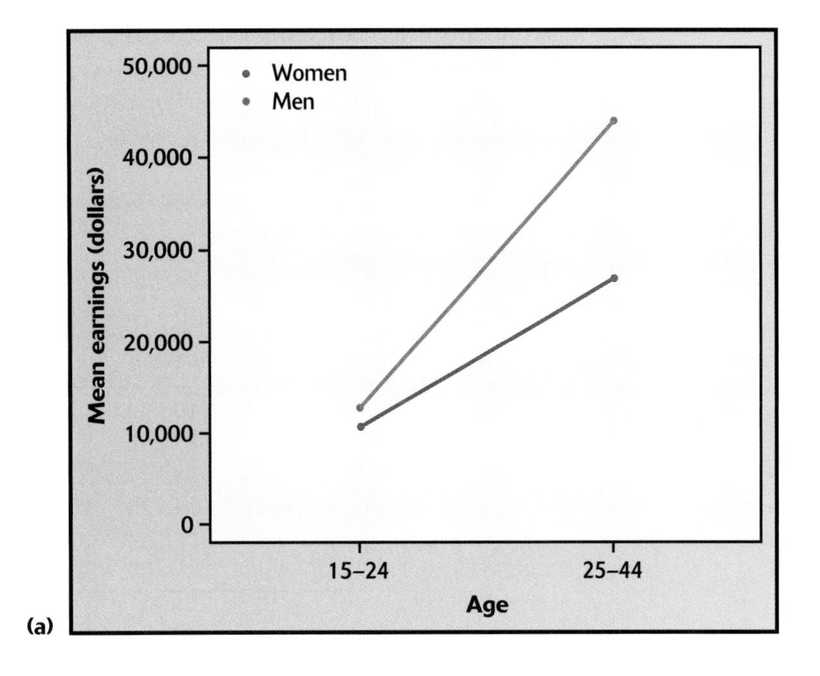

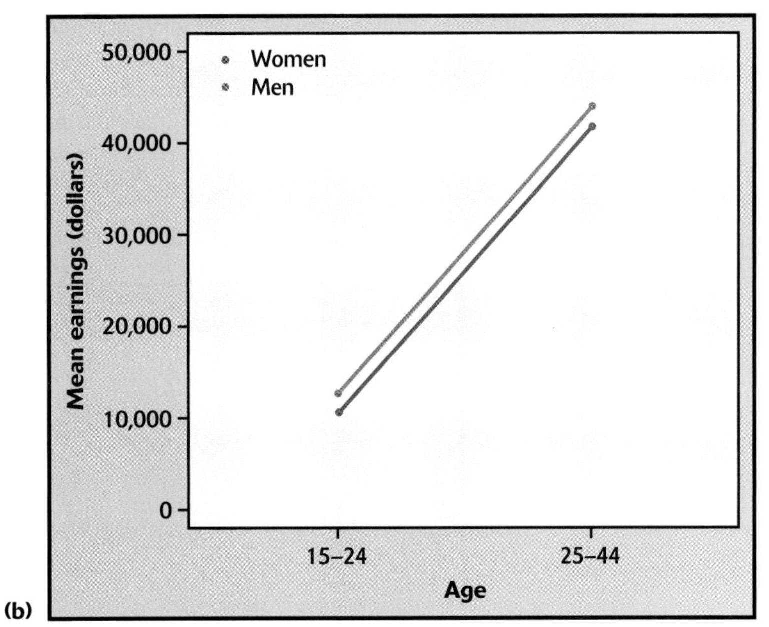

FIGURE 15.1 (a) Plot of mean earnings of men and women in two age groups, for Example 15.4. Interaction between age and gender is visible in the lack of parallelism of the lines. (b) The plot as it would appear if there were no interaction. The male and female lines are now parallel.

Figure 15.1(a) is a plot of the four group means. Because the difference between the mean earnings of men and women increases with age, the gap between the lines increases from left to right. That is, the male and female lines are not parallel.

How would the plot look if there were no interaction? No interaction says that the male-female earnings gap is the same in both age groups. That is, the effect of gender does not depend on age. Suppose that the gap were $2434 in both age groups. Figure 15.1(b) plots the means. The male and female lines are parallel. *Interaction between the factors is visible as lack of parallelism in a plot of the group means.*

EXAMPLE 15.5 **Workers' earnings, continued**

The Bureau of Labor Statistics in fact reports mean earnings for four age groups. Here are the data:

Age	Women	Men	Mean
15–24	$10,744	$13,178	$11,961
25–44	$27,248	$44,224	$35,736
45–64	$28,771	$54,487	$41,629
65 and over	$14,924	$32,819	$23,872
Mean	$20,422	$36,177	$28,299

Including the additional age groups changes the marginal means for gender and the overall mean of all groups. Figure 15.2 is a plot of the group means. The plot shows a clear main effect for gender: women earn less than men in every age group. There is also a clear main effect for age: the mean earnings of both men and women increase with age as we move through the first three age groups, then decrease in the 65 and over group. The male and female lines are not parallel, so interaction is present. *The main effects are meaningful and important despite the interaction.* We will see that this is not always true.

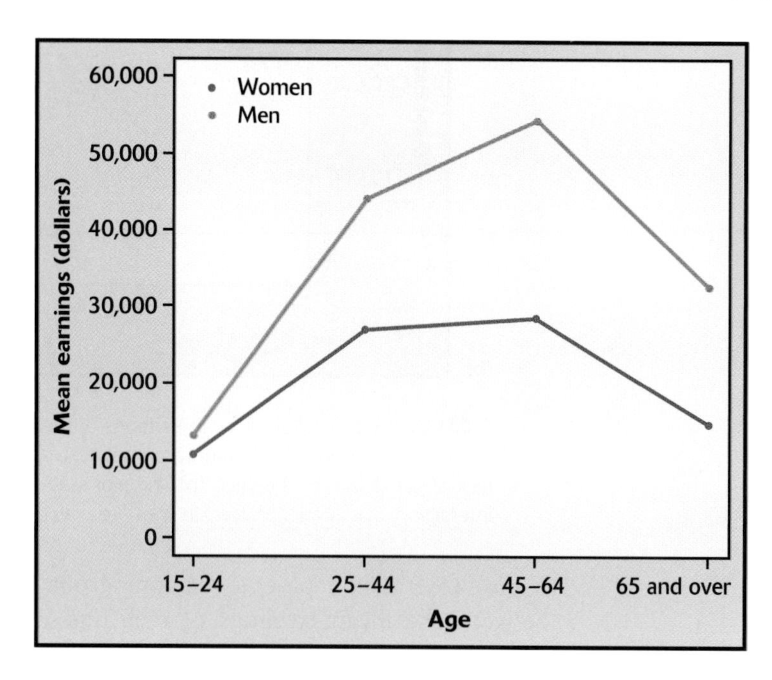

FIGURE 15.2 Plot of mean earnings of men and women in four age groups, for Example 15.5. The plot shows clear main effects for both age and gender as well as interaction between the two factors.

15.4 **Marginal means.** Verify the marginal mean for men given in Example 15.5. Then verify that the overall mean at the lower right of the table is the average of the two gender means and also the average of the four age group means.

15.5 **How do the differences depend on age?** One way to describe the interaction between age and gender in Example 15.5 is to give the differences between the earnings of men and the earnings of women in the four age groups. Plot the differences versus age and write a short summary of what you conclude from the plot.

15.6 **Lack of interaction.** Suppose that the difference between the mean earnings of men and women remained fixed at $2434 for all four age groups in Example 15.5, and that the mean earnings for men in each age group are as given in the table. Find the mean earnings for women in each age group and make a plot of the 8 group means. In what important way does your plot differ from Figure 15.2?

Interactions come in many forms. When we find them, a careful examination of the means is needed to properly interpret the data. Simply stating that interactions are significant tells us little. Plots of the group means are very helpful.

EXAMPLE 15.6 **Energy expenditure of African farmers**

A study of the energy expenditure of farmers in the African nation of Burkina Faso collected data during the wet and dry seasons.[4] The farmers grow millet during the wet season. In the dry season, there is relatively little activity because the ground is too hard to grow crops. Here is the mean energy expended (in calories) by men and women in Burkina Faso during the wet and dry seasons:

Season	Men	Women	Mean
Dry	2310	2320	2315
Wet	3460	2890	3175
Mean	2885	2605	2745

Figure 15.3 is the plot of the group means.

During the dry season both men and women use about the same number of calories. When the wet season arrives, both genders use more energy. The men, who by social convention do most of the field work, expend much more energy than the women. The amount of energy used by the men in the wet season is very high by any reasonable standard. Such values are sometimes found in developed countries for people engaged in coal mining and lumberjacking.

In a statistical analysis, the pattern of means shown in Figure 15.3 produced significant main effects for season and gender in addition to a season-by-gender interaction. The main effects record that men use more energy than women and that the wet season is associated with greater activity than the dry season. This clearly does not tell the whole story. We need to

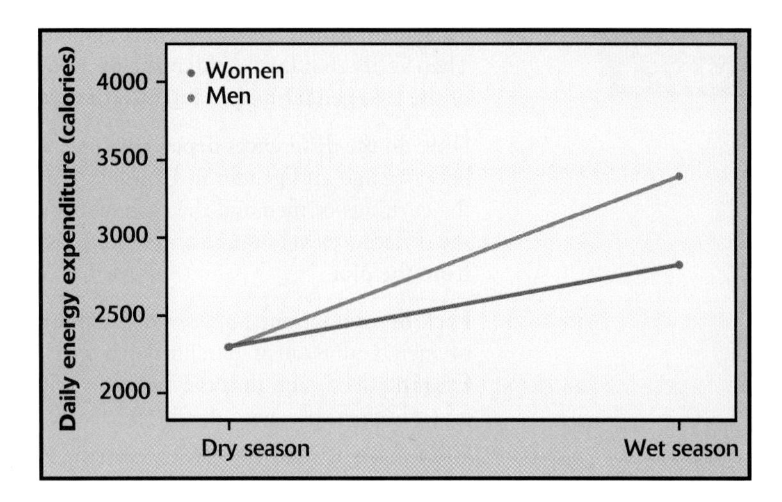

FIGURE 15.3 Plot of mean energy expenditures of farmers in Burkina Faso by gender and season, for Example 15.6.

discuss the men and women in each of the two seasons to fully understand how energy is used in Burkina Faso.

A different kind of interaction is present in the next example. Here, we must be very cautious in our interpretation of the main effects since one of them can lead to a distorted conclusion.

EXAMPLE 15.7 **Noise and working on math problems**

An experiment to study how noise affects the performance of children tested second-grade hyperactive children and a control group of second graders who were not hyperactive.[5] One of the tasks involved solving math problems. The children solved problems under both high-noise and low-noise conditions. Here are the mean scores:

Group	High noise	Low noise	Mean
Control	214	170	192
Hyperactive	120	140	130
Mean	167	155	161

The means are plotted in Figure 15.4. In the analysis of this experiment, both of the main effects and the interaction were statistically significant. How are we to interpret these results?

What catches our eye in the plot is that the lines cross. The hyperactive children did better in low noise than in high, but the control children performed better in high noise than in low. This interaction between noise and type of children is an important result. Which noise level is preferable depends on which type of children are in the class. The main effect for type of children is straightforward: the normal children did better than the

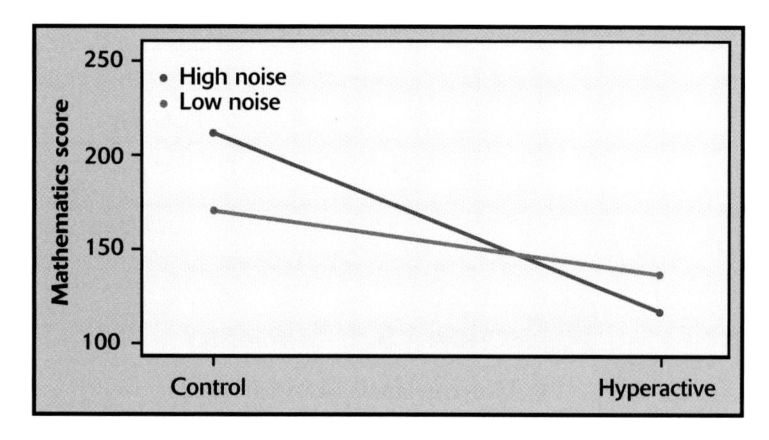

FIGURE 15.4 Plot of mean math scores of control and hyperactive children under low- and high-noise conditions, for Example 15.7.

hyperactive children at both levels of noise. However, *the main effect for noise is not practically meaningful because of the interaction.* Under high-noise conditions the marginal mean was 167, compared with a low-noise mean of 155. This might suggest that high noise is, on the average, a more favorable condition than low noise. In fact, high noise is more favorable only for the control group. The hyperactive children performed better under the low-noise condition. Because of the interaction, we must know which type of children we have in order to say which noise level is better for working math problems.

APPLY YOUR KNOWLEDGE

15.7 **Is there an interaction?** Each of the following tables gives means for a two-way ANOVA. Make a plot of the means with the levels of factor A on the x axis. State whether or not there is an interaction, and if there is, describe it.

(a)

	Factor A		
Factor B	1	2	3
1	10	20	30
2	20	40	60

(b)

	Factor A		
Factor B	1	2	3
1	10	20	30
2	30	20	10

(c)

	Factor A		
Factor B	1	2	3
1	10	20	30
2	20	30	40

(d)

	Factor A		
Factor B	1	2	3
1	10	20	30
2	10	20	60

Section 15.1 Summary

■ **Two-way analysis of variance** is used to compare population means when populations are classified according to two factors.

■ ANOVA assumes that the populations are Normal and that independent SRSs are drawn from each population. The populations may have different means, but ANOVA assumes that they have the same standard deviation.

■ As with one-way ANOVA, preliminary analysis includes examination of means, standard deviations, and perhaps Normal quantile plots. **Marginal means** are calculated by taking averages of the group means across rows and columns. Pooling is used to estimate the within-group variance.

■ ANOVA separates the total variation into parts for the **model** and **error.** The model variation is separated into parts for each of the **main effects** and the **interaction.**

15.2 Inference for Two-Way ANOVA

Because two-way ANOVA breaks the FIT part of the model into three parts, corresponding to the two main effects and the interaction, inference for two-way ANOVA includes an F statistic for each of these effects. As with one-way ANOVA, the calculations are organized in an ANOVA table.

The ANOVA table for two-way ANOVA

The results of a two-way ANOVA are summarized in an ANOVA table based on splitting the total variation SST and the total degrees of freedom DFT among the two main effects and the interaction. *When the sample size is the same for all groups,* both the sums of squares (which measure variation) and the degrees of freedom add:

$$\text{SST} = \text{SSA} + \text{SSB} + \text{SSAB} + \text{SSE}$$
$$\text{DFT} = \text{DFA} + \text{DFB} + \text{DFAB} + \text{DFE}$$

When the n_{ij} are not all equal, there are several ways to decompose SST, and the sums of squares may not add. Whenever possible, design studies with equal sample sizes to avoid these complications. We will do inference only for the equal sample size case.

The sums of squares are always calculated in practice by statistical software. From each sum of squares and its degrees of freedom we find the mean square in the usual way:

$$\text{mean square} = \frac{\text{sum of squares}}{\text{degrees of freedom}}$$

The significance of each of the main effects and the interaction is assessed by an F statistic that compares the variation due to the effect of interest with the within-group variation. Each F statistic is the mean square for the source of interest divided by MSE. Here is the general form of the two-way ANOVA table:

Source	Degrees of freedom	Sum of squares	Mean square	F
A	$I - 1$	SSA	SSA/DFA	MSA/MSE
B	$J - 1$	SSB	SSB/DFB	MSB/MSE
AB	$(I - 1)(J - 1)$	SSAB	SSAB/DFAB	MSAB/MSE
Error	$N - IJ$	SSE	SSE/DFE	
Total	$N - 1$	SST	SST/DFT	

There are three null hypotheses in two-way ANOVA, with an F test for each. We can test for significance of the main effect of A, the main effect of B, and the AB interaction. It is generally good practice to examine the test for interaction first, since the presence of a strong interaction may influence the interpretation of the main effects. Be sure to plot the means as an aid to interpreting the results of the significance tests.

SIGNIFICANCE TESTS IN TWO-WAY ANOVA

To test the main effect of A, use the F statistic

$$F_A = \frac{\text{MSA}}{\text{MSE}}$$

To test the main effect of B, use the F statistic

$$F_B = \frac{\text{MSB}}{\text{MSE}}$$

To test the interaction of A and B, use the F statistic

$$F_{AB} = \frac{\text{MSAB}}{\text{MSE}}$$

If the effect being tested is zero, the calculated F statistic has an F distribution with numerator degrees of freedom corresponding to the effect and denominator degrees of freedom equal to DFE. Large values of the F statistic lead to rejection of the null hypothesis. The P-value is the probability that a random variable having the corresponding F distribution is greater than or equal to the calculated value.

15.8 **Comparing employee training programs.** Exercise 15.1 (page 15-8) describes the setting for a two-way ANOVA design that compares employee training programs. Give the degrees of freedom for each of the F statistics that are used to test the main effects and the interaction for this problem.

15.9 **Customers' preferences for packaging.** Exercise 15.2 (page 15-9) describes the setting for a two-way ANOVA design that compares different packages for computer disks. Give the degrees of freedom for each of the F statistics that are used to test the main effects and the interaction for this problem.

15.10 **Hand lotion marketing.** Exercise 15.3 (page 15-9) describes the setting for a two-way ANOVA design that compares different hand lotion formulations. Give the degrees of freedom for each of the F statistics that are used to test the main effects and the interaction for this problem.

Carrying out a two-way ANOVA

The following example illustrates how to do a two-way ANOVA. As with the one-way ANOVA, we focus our attention on interpretation of the computer output.

CASE 15.1

DISCOUNTS AND EXPECTED PRICES

Does the frequency with which a supermarket product is offered at a discount affect the price that customers expect to pay for the product? Does the percent reduction also affect this expectation? These questions were examined by researchers in a study conducted on students enrolled in an introductory management course at a large midwestern university. For 10 weeks 160 subjects received information about the products. The treatment conditions corresponded to the number of promotions (1, 3, 5, or 7) during this 10-week period and the percent that the product was discounted (10%, 20%, 30%, and 40%). Ten students were randomly assigned to each of the $4 \times 4 = 16$ treatments.[6] For our case study we will examine the data for two levels of promotions (1 and 3) and two levels of discount (40% and 20%). Thus, we have a two-way ANOVA with each of the factors having two levels and 10 observations in each of the 4 treatment combinations. Here are the data:

Number of promotions	Percent discount	Expected price									
1	40	4.10	4.50	4.47	4.42	4.56	4.69	4.42	4.17	4.31	4.59
1	20	4.94	4.59	4.58	4.48	4.55	4.53	4.59	4.66	4.73	5.24
3	40	4.07	4.13	4.25	4.23	4.57	4.33	4.17	4.47	4.60	4.02
3	20	4.88	4.80	4.46	4.73	3.96	4.42	4.30	4.68	4.45	4.56

As usual we start our statistical analysis with a careful examination of the data.

EXAMPLE 15.8 **Plotting the data**

CASE 15.1

With 10 observations per treatment, we can plot the individual observations. To do this we created an additional variable, "Comb," that has four distinct values corresponding to the particular combination of the number of promotions and the discount. The value "d20-p1" corresponds to 20% discount with 1 promotion, and the values "d20-p3," "d40-p1," and "d40-p3" have similar interpretations. The data are plotted in Figure 15.5. The lines in the figure connect the four group means.

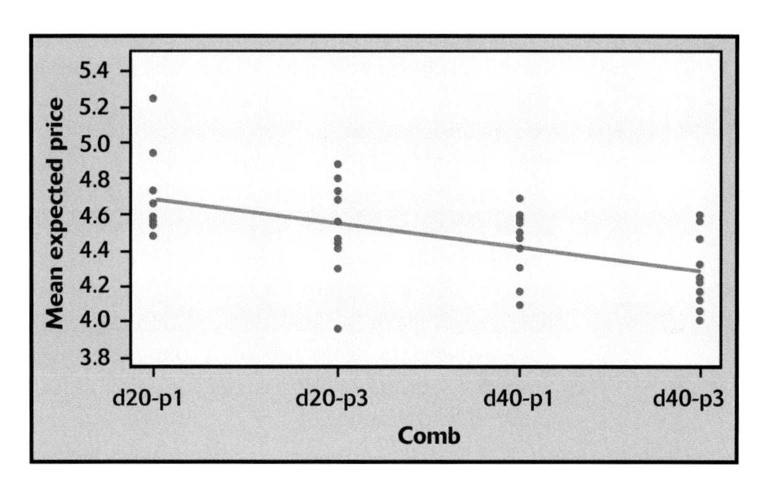

FIGURE 15.5 Plot of the data for the promotions and discount example, Example 15.8. The four treatments appear along the horizontal axis. The lines connect the group means.

The spread of the data within each group is similar and there are no outliers or other unusual patterns. That is, the conditions for ANOVA inference appear to be satisfied. The treatment means appear to differ.

CASE 15.2

EXPECTED PRICES, CONTINUED

Our second case study is a variation on the first. We will use data from the experiment described in Case 15.1 but with different treatment combinations. Here are the data for the factor promotions at levels 1 and 5, and the factor discount at levels 30% and 10%:

Number of promotions	Percent discount	Expected price									
1	30	3.57	3.77	3.90	4.49	4.00	4.66	4.48	4.64	4.31	4.43
1	10	5.19	4.88	4.78	4.89	4.69	4.96	5.00	4.93	5.10	4.78
5	30	3.90	3.77	3.86	4.10	4.10	3.81	3.97	3.67	4.05	3.67
5	10	4.31	4.36	4.75	4.62	3.74	4.34	4.52	4.37	4.40	4.52

15.11 Plot the data for Case 15.2. Make a plot similar to the one given in Figure 15.5 for the levels of the factors given in Case 15.2. Connect the means with lines. Do the conditions for ANOVA inference appear to be met? Describe the pattern of the group means.

After looking at the data graphically, we proceed with numerical summaries.

EXAMPLE 15.9 **Means and standard deviations**

The software output in Figure 15.6 gives descriptive statistics for the data of Case 15.1. In the row with 1 under the heading "PROMO" and 20 under the heading "DISCOUNT," the mean of the 10 observations in this treatment combination is given as 4.689. The standard deviation is 0.2331. We would report these as 4.69 and 0.23. The next row gives results for 1 promotion and a 40% discount. The marginal results for all 20 students assigned to 1 promotion appear in the following "Total" row. The marginal standard deviation 0.2460 is not useful because it ignores the fact that the 10 observations for the 20% discount and the 10 observations for the 40% discount come from different populations. The overall mean for all 40 observations appears in the last row of the table. The standard deviations for the four groups are quite similar, and we have no reason to suspect a serious violation of the condition that the population standard deviations must all be the same.

Descriptive Statistics

Dependent Variable: EPRICE

PROMO	DISCOUNT	Mean	Std. Deviation	N
1	20	4.689	0.2331	10
	40	4.423	0.1848	10
	Total	4.556	0.2460	20
3	20	4.524	0.2707	10
	40	4.284	0.2040	10
	Total	4.404	0.2638	20
Total	20	4.607	0.2600	20
	40	4.353	0.2024	20
	Total	4.480	0.2633	40

FIGURE 15.6 Descriptive statistics from SPSS for the promotions and discount example, for Example 15.9.

Often we display the means in a table similar to the following:

	Discount		
Promotions	20%	40%	Total
1	4.69	4.42	4.56
3	4.52	4.28	4.40
Total	4.61	4.35	4.48

In this table the marginal means give us information about the main effects. When promotions are increased from 1 to 3, the expected price drops from \$4.56 to \$4.40. Furthermore, when the discount is increased from 20% to 40%, the expected price drops from \$4.61 to \$4.35.

Numerical summaries with marginal means enable us to describe the main effects in a two-way ANOVA. For interactions, however, graphs are much better.

EXAMPLE 15.10 **Plotting the group means**

The means for the promotions and discount data of Case 15.1 are plotted in Figure 15.7. We have chosen to put the two values of promotion on the x axis. We see that the mean expected price for the 40% discount condition is consistently lower than the mean expected price for the 20% discount condition. Similarly, the means for 3 promotions are consistently less than the means for 1 promotion. The two lines are approximately parallel, suggesting that there is little interaction between promotion and discount in this example.

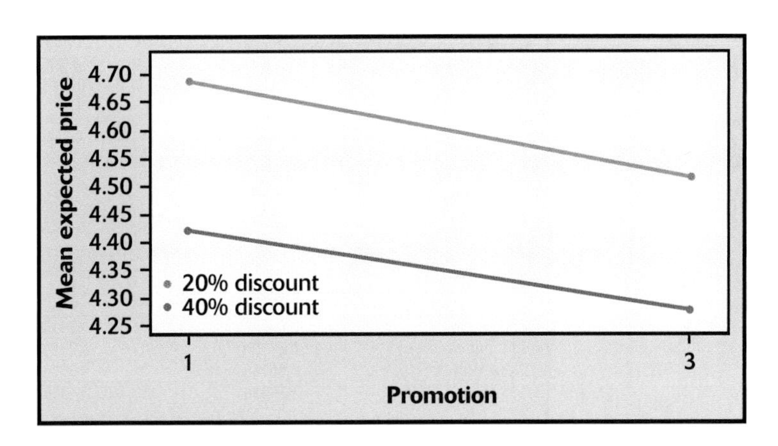

FIGURE 15.7 Plot of the means for the promotions and discount example, for Example 15.10.

15.12 **Group means in Excel output.** The first part of the Excel output in Figure 15.8 gives the group means and the marginal means for the data in Case 15.1. Find these means and display them in a table. Report them with the digits exactly as given in the output. Does Excel agree with the SPSS output in Figure 15.6? (CASE 15.1)

15.13 **Numerical summaries for Case 15.2.** Find the means and standard deviations for each of the promotion-by-discount treatment combinations for the data in Case 15.2. Display the means in a table that also includes the marginal means. Plot the means and describe the main effects and the interaction. Do the standard deviations suggest that it is reasonable to pool the group standard deviations to get MSE? (CASE 15.2)

Minitab

Analysis of Variance for Eprice

Source	DF	SS	MS	F	P
Promo	1	0.2310	0.2310	4.54	0.040
Discount	1	0.6401	0.6401	12.59	0.001
Interaction	1	0.0017	0.0017	0.03	0.856
Error	36	1.8304	0.0508		
Total	39	2.7032			

Microsoft Excel - promo.xls

File Edit View Insert Format Tools Data Window Help

	A	B	C	D	E	F	G
1	Anova: Two-Factor With Replication						
2							
3	SUMMARY	1	3	Total			
4	*40*						
5	Count	10	10	20			
6	Sum	44.23	42.84	87.07			
7	Average	4.423	4.284	4.3535			
8	Variance	0.034134444	0.041626667	0.040971316			
9							
10	*20*						
11	Count	10	10	20			
12	Sum	46.89	45.24	92.13			
13	Average	4.689	4.524	4.6065			
14	Variance	0.054321111	0.073293333	0.067613421			
15							
16	*Total*						
17	Count	20	20				
18	Sum	91.12	88.08				
19	Average	4.556	4.404				
20	Variance	0.06052	0.069593684				
21							
22							
23	ANOVA						
24	*Source of Variation*	*SS*	*df*	*MS*	*F*	*P-value*	*F crit*
25	Sample	0.64009	1	0.64009	12.58932025	0.001100288	4.113161367
26	Columns	0.23104	1	0.23104	4.544105596	0.03992794	4.113161367
27	Interaction	0.00169	1	0.00169	0.033239	0.856357802	4.113161367
28	Within	1.83038	36	0.050843889			
29							
30	Total	2.7032	39				

Sheet1 / A2 /

Ready NUM

FIGURE 15.8 Two-way ANOVA output from Minitab and Excel for the promotions and discount example, for Exercise 15.12 and Example 15.10. *(continued)*

Having examined the data carefully using numerical and graphical summaries, we are now ready to proceed with the statistical examination of the data using the two-way ANOVA model.

EXAMPLE 15.11 **ANOVA software output**

Figure 15.8 gives the two-way ANOVA output from Minitab, Excel, SPSS, and SAS. Look first at the ANOVA table in the Minitab output. The form of the table is very similar to the general form of the two-way ANOVA table given on page 15-17. In place of A and B as the generic factors, the output gives the labels that we specified

SPSS

Descriptive Statistics
Dependent Variable: EPRICE

PROMO	DISCOUNT	Mean	Std. Deviation	N
1	20	4.689	0.2331	10
	40	4.423	0.1848	10
	Total	4.556	0.2460	20
3	20	4.524	0.2707	10
	40	4.284	0.2040	10
	Total	4.404	0.2638	20
Total	20	4.607	0.2600	20
	40	4.353	0.2024	20
	Total	4.480	0.2633	40

Tests of Between-Subjects Effects
Dependent Variable: EPRICE

Source	Type III Sum of Squares	df	Mean Square	F	Sig.
Corrected Model	0.873	3	0.291	5.722	0.003
Intercept	802.816	1	802.816	15789.823	0.000
PROMO	0.231	1	0.231	4.544	0.040
DISCOUNT	0.640	1	0.640	12.589	0.001
PROMO * DISCOUNT	1.690E-03	1	1.690E-03	0.033	0.856
Error	1.830	36	5.084E-02		
Total	805.519	40			
Corrected Total	2.703	39			

a R Squared = 0.323 (Adjusted R Squared = 0.266)

SAS

The GLM Procedure
Dependent Variable: Eprice

Source	DF	Sum of Squares	Mean Square	F Value	Pr > F
Model	3	0.87282000	0.29094000	5.72	0.0026
Error	36	1.83038000	0.05084389		
Corrected Total	39	2.70320000			

R-Squared	Coeff Var	Root MSE	Eprice Mean
0.322884	5.033167	0.225486	4.480000

Source	DF	Type I SS	Mean Square	F Value	Pr > F
Promo	1	0.23104000	0.23104000	4.54	0.0399
Discount	1	0.64009000	0.64009000	12.59	0.0011
Promo*Discount	1	0.00169000	0.00169000	0.03	0.8564

FIGURE 15.8 *(continued)* Two-way ANOVA output from SPSS and SAS for the promotions and discount example, for Exercise 15.12 and Example 15.10.

when we entered the data. We have main effects for "Discount" and "Promo." The interaction between these two factors is simply labeled "Interaction," and the last two rows are "Error" and "Total." The results of the significance tests are in the last two columns, labeled "F" and "P." As expected, the interaction is not statistically significant ($F = 0.03$, df $= 1$ and 36, $P = 0.856$). On the other hand, the main effects of discount ($F = 12.59$, df $= 1$ and 36, $P = 0.001$) and promotion ($F = 4.54$, df $= 1$ and 36, $P = 0.040$) are significant.

The statistical significance tests assure us that the differences that we saw in the graphical and numerical summaries can be distinguished from chance variation. We summarize as follows: When promotions are increased from 1 to 3, the expected price drops from \$4.56 to \$4.40 ($F = 4.54$, df = 1 and 36, $P = 0.040$). Furthermore, when the discount is increased from 20% to 40%, the expected price decreases from \$4.61 to \$4.35 ($F = 12.59$, df = 1 and 36, $P = 0.001$).

Minitab does not explicitly give s, the estimate of the parameter σ of our model. To find this, we take the square root of the mean square error, $s = \sqrt{0.0508} = 0.225$.

APPLY YOUR KNOWLEDGE

15.14 Verify the ANOVA calculations. Use the output in Figure 15.8 to verify that the four mean squares are obtained by dividing the corresponding sums of squares by the degrees of freedom. Similarly, show how each F statistic is obtained by dividing two of the mean squares.

15.15 Compare software outputs. Examine the outputs from Minitab, Excel, SPSS, and SAS carefully. Write a short evaluation comparing the formats. Indicate which you prefer and why.

15.16 Run the ANOVA for Case 15.2. Analyze the data for Case 15.2 using the two-way ANOVA model and summarize the results.

SECTION 15.2 SUMMARY

■ The calculations for two-way ANOVA are organized in an **ANOVA table.**

■ F statistics and P-values are used to test hypotheses about the main effects and the interaction.

■ Careful inspection of the means is necessary to interpret significant main effects and interactions. Plots are a useful aid.

STATISTICS IN SUMMARY

Two-way ANOVA compares populations that are classified according to two categorical explanatory variables. Many of the basic ideas are similar to those that we learned in the previous chapter on one-way ANOVA. However, after studying this chapter, you should be able to do the following.

A. RECOGNITION

1. Recognize when a two-way ANOVA analysis is helpful in understanding data.
2. Recognize that the statistical significance in two-way ANOVA depends on the sizes of the samples and on how much variation there is within the samples.
3. Recognize when you can safely use two-way ANOVA. Check the data production, the presence of outliers, and the sample standard deviations for each combination of factor levels.

B. INTERPRETING ANOVA

1. Use graphical and numerical summaries to describe main effects and interaction.
2. Explain the null hypotheses for main effects and interactions in a specific setting.
3. Locate the F statistics and P-values on the output of a computer two-way ANOVA program.
4. Find the degrees of freedom for the F statistics from the number of levels for each of the two factors and the number of observations. Use the F critical values in Table E to approximate the P-values.

CHAPTER 15 REVIEW EXERCISES

15.17 Describe the design. Each of the following situations is a two-way study design. For each case, identify the response variable and both factors, and state the number of levels for each factor (I and J) and the total number of observations (N).

(a) A study of smoking classifies subjects as nonsmokers, moderate smokers, or heavy smokers. Samples of 80 men and 80 women are drawn from each group. Each person reports the number of hours of sleep he or she gets on a typical night.

(b) The strength of concrete depends upon the formula used to prepare it. An experiment compares six different mixtures. Nine specimens of concrete are poured from each mixture. Three of these specimens are subjected to 0 cycles of freezing and thawing, three are subjected to 100 cycles, and three specimens are subjected to 500 cycles. The strength of each specimen is then measured.

(c) Four methods for teaching sign language are to be compared. Sixteen students in special education and sixteen students majoring in other areas are the subjects for the study. Within each group they are randomly assigned to the methods. Scores on a final exam are compared.

15.18 Outline the ANOVA table. For each part of Exercise 15.17, outline the ANOVA table, giving the sources of variation and the degrees of freedom. (Do not compute the numerical values for the sums of squares and mean squares.)

15.19 Degrees of freedom and the critical value. A two-way ANOVA model was used to analyze an experiment with three levels of one factor, four levels of a second factor, and 6 observations per treatment combination.

(a) For each of the main effects and the interaction, give the degrees of freedom for the corresponding F statistic.

(b) Using Table E or statistical software, find the value that each of these F statistics must exceed for the result to be significant at the 5% level.

(c) Answer part (b) for the 1% level.

15.20 Degrees of freedom and the critical value. A two-way ANOVA model was used to analyze an experiment with two levels of one factor, three levels of a second factor, and 6 observations per treatment combination.

(a) For each of the main effects and the interaction, give the degrees of freedom for the corresponding F statistic.

(b) Using Table E or statistical software, find the value that each of these F statistics must exceed for the result to be significant at the 5% level.

(c) Answer part (b) for the 1% level.

15.21 Systolic blood pressure. In the course of a clinical trial of measures to prevent coronary heart disease, blood pressure measurements were taken on 12,866 men. Individuals were classified by age group and race.[7] The means for systolic blood pressure are given in the following table:

	35–39	40–44	45–49	50–54	55–59
White	131.0	132.3	135.2	139.4	142.0
Nonwhite	132.3	134.2	137.2	141.3	144.1

(a) Plot the group means, with age on the x axis and blood pressure on the y axis. For each racial group connect the points for the different ages.

(b) Describe the patterns you see. Does there appear to be a difference between the two racial groups? Does systolic blood pressure appear to vary with age? If so, how does it vary? Is there an interaction?

(c) Compute the marginal means. Then find the differences between the white and nonwhite mean blood pressures for each age group. Use this information to summarize numerically the patterns in the plot.

15.22 Diastolic blood pressure. The means for diastolic blood pressure recorded in the clinical trial described in the previous exercise are:

	35–39	40–44	45–49	50–54	55–59
White	89.4	90.2	90.9	91.6	91.4
Nonwhite	91.2	93.1	93.3	94.5	93.5

(a) Plot the group means with age on the x axis and blood pressure on the y axis. For each racial group connect the points for the different ages.

(b) Describe the patterns you see. Does there appear to be a difference between the two racial groups? Does diastolic blood pressure appear to vary with age? If so, how does it vary? Is there an interaction between race and age?

(c) Compute the marginal means. Find the differences between the white and nonwhite mean blood pressures for each age group. Use this information to summarize numerically the patterns in the plot.

15.23 Chromium and insulin. The amount of chromium in the diet has an effect on the way the body processes insulin. In an experiment designed to study this phenomenon, four diets were fed to male rats. There were two factors. Chromium had two levels: low (L) and normal (N). The rats were allowed to eat as much as they wanted (M) or the total amount that they could eat was restricted (R). We call the second factor Eat. One of the variables measured was the amount of an enzyme called GITH.[8] The means for this response variable appear in the following table:

	Eat	
Chromium	M	R
L	4.545	5.175
N	4.425	5.317

(a) Make a plot of the mean GITH for these diets, with the factor Chromium on the x axis and GITH on the y axis. For each Eat group connect the points for the two Chromium means.

(b) Describe the patterns you see. Does the amount of chromium in the diet appear to affect the GITH mean? Does restricting the diet rather than letting the rats eat as much as they want appear to have an effect? Is there an interaction?

(c) Compute the marginal means. Compute the differences between the M and R diets for each level of Chromium. Use this information to summarize numerically the patterns in the plot.

15.24 Gender, majors, and social insight. The Chapin Social Insight Test measures how well people can appraise others and predict what they may say or do. A study administered this test to different groups of people and compared the mean scores.[9] Some of the results are given in the table below. Means for males and females who were psychology graduate students (PG) and liberal arts undergraduates (LU) are presented. The two factors are labeled Gender and Group.

	Group	
Gender	PG	LU
Males	27.56	25.34
Females	29.25	24.94

Plot the means and describe the essential features of the data in terms of main effects and interactions.

15.25 Gender, changing majors, and SAT math scores. The change-of-majors study described in Example 14.3 (page 14-8) classified students into one of three groups depending upon their major in the sophomore year. In this exercise we also consider the gender of the students. There are now two factors: major with three levels, and gender with two levels. The mean SAT mathematics scores for the six groups appear in the following table. For convenience we use the labels CS for computer science majors, EO for engineering and other science majors, and O for other majors.

	Major		
Gender	CS	EO	O
Males	628	618	589
Females	582	631	543

Describe the main effects and interaction using appropriate graphs and calculations.

15.26 Gender, changing majors, and high school math grades. The mean high school mathematics grades for the students in the previous exercise are summarized in the following table. The grades have been coded so that 10 = A, 9 = A−, etc.

Gender	Major CS	EO	O
Males	8.68	8.35	7.65
Females	9.11	9.36	8.04

Summarize the results of this study using appropriate plots and calculations to describe the main effects and interaction.

15.27 **A new material to repair wounds.** One way to repair serious wounds is to insert some material as a scaffold for the body's repair cells to use as a template for new tissue. Scaffolds made from extracellular material (ECM) are particularly promising for this purpose. Because ECMs are biological material, they serve as an effective scaffold and are then absorbed. Unlike biological material that includes cells, however, they do not trigger tissue rejection reactions in the body. One study compared 6 types of scaffold material.[10] Three of these were ECMs and the other three were made of inert materials. There were three mice used per scaffold type. The response measure was the percent of glucose phosphated isomerase (Gpi) cells in the region of the wound. A large value is good, indicating that there are many bone marrow cells sent by the body to repair the tissue. In Exercise 14.58 (page 14-50) we analyzed the data for rats whose tissues were measured 4 weeks after the repair. The experiment included additional groups of rats who received the same types of scaffold but were measured at different times. Here are the data for 4 weeks and 8 weeks after the repair:

Material	Gpi (%) 4 weeks			8 weeks		
ECM1	55	70	70	60	65	65
ECM2	60	65	65	60	70	60
ECM3	75	70	75	70	80	70
MAT1	20	25	25	15	25	25
MAT2	5	10	5	10	5	5
MAT3	10	15	10	5	15	10

(a) Make a table giving the sample size, mean, and standard deviation for each of the material-by-time combinations. Is it reasonable to pool the variances? Because the sample sizes in this experiment are very small, we expect a large amount of variability in the sample standard deviations. Although they vary more than we would prefer, we will proceed with the ANOVA.

(b) Make a plot of the means. Describe the main features of the plot.

(c) Run the analysis of variance. Report the F statistics with degrees of freedom and P-values for each of the main effects and the interaction. What do you conclude? Write a short paragraph summarizing the results of your analysis.

15.28 **Add an additional level for a factor.** Refer to the previous exercise. Here are the data that were collected at 2 weeks, 4 weeks, and 8 weeks:

Material	Gpi (%)								
	2 weeks			4 weeks			8 weeks		
ECM1	70	75	65	55	70	70	60	65	65
ECM2	60	65	70	60	65	65	60	70	60
ECM3	80	60	75	75	70	75	70	80	70
MAT1	50	45	50	20	25	25	15	25	25
MAT2	5	10	15	5	10	5	10	5	5
MAT3	30	25	25	10	15	10	5	15	10

Rerun the analyses that you performed in the previous exercise. How does the addition of the data for 2 weeks change the conclusions? Write a summary comparing these results with those in the previous exercise.

15.29 **Analyze the results for each time period.** Refer to the previous exercise. Analyze the data for each time period separately using a one-way ANOVA. Use a multiple comparisons procedure where needed. Summarize the results.

15.30 **Cooking pots and dietary iron.** Iron-deficiency anemia is the most common form of malnutrition in developing countries, affecting about 50% of children and women and 25% of men. Iron pots for cooking foods had traditionally been used in many of these countries, but they have been largely replaced by aluminum pots, which are cheaper and lighter. Some research has suggested that food cooked in iron pots will contain more iron than food cooked in other types of pots. One study designed to investigate this issue compared the iron content of some Ethiopian foods cooked in aluminum, clay, and iron pots.[11] In Exercise 14.56, we analyzed the iron content of *yesiga wet'*, beef cut into small pieces and prepared with several Ethiopian spices. The researchers who conducted this study also examined the iron content of *shiro wet'*, a legume-based mixture of chickpea flour and Ethiopian spiced pepper, and *ye-atkilt allych'a,* a lightly spiced vegetable casserole. In the table below, these three foods are labeled meat, legumes, and vegetables. Four samples of each food were cooked in each type of pot. The iron in the food is measured in milligrams of iron per 100 grams of cooked food. Here are the data:

Type of pot	Iron content											
	Meat				Legumes				Vegetables			
Aluminum	1.77	2.36	1.96	2.14	2.40	2.17	2.41	2.34	1.03	1.53	1.07	1.30
Clay	2.27	1.28	2.48	2.68	2.41	2.43	2.57	2.48	1.55	0.79	1.68	1.82
Iron	5.27	5.17	4.06	4.22	3.69	3.43	3.84	3.72	2.45	2.99	2.80	2.92

(a) Make a table giving the sample size, mean, and standard deviation for each type of pot. Is it reasonable to pool the variances? Although the standard deviations vary more than we would like, this is partially due to the small sample sizes and we will proceed with the analysis of variance.

(b) Plot the means. Give a short summary of how the iron content of foods depends upon the cooking pot.

(c) Run the analysis of variance. Give the ANOVA table, the F statistics with degrees of freedom and P-values, and your conclusions regarding the hypotheses about main effects and interactions.

15.31 Main effects versus interaction. Refer to the previous exercise. Although there is a statistically significant interaction, do you think that these data support the conclusion that foods cooked in iron pots contain more iron than foods cooked in aluminum or clay pots? Discuss.

15.32 Rerun as a one-way ANOVA. Refer to Exercise 15.30. Rerun the analysis as a one-way ANOVA with 9 groups and 4 observations per group. Report the results of the F test. Examine differences in means using a multiple comparisons procedure. Summarize your results and compare them with those you obtained in Exercise 15.30.

15.33 A manufacturing problem. One step in the manufacture of large engines requires that holes of very precise dimensions be drilled. The tools that do the drilling are regularly examined and are adjusted to ensure that the holes meet the required specifications. Part of the examination involves measurement of the diameter of the drilling tool. A team studying the variation in the sizes of the drilled holes selected this measurement procedure as a possible cause of variation in the drilled holes. They decided to use a designed experiment as one part of this examination. Some of the data are given in Table 15.1. The diameters in millimeters (mm) of five tools were measured by the same operator at three times (8:00 A.M., 11:00 A.M., and 3:00 P.M.). Three measurements were taken on each tool at each time. The person taking the measurements could not tell which tool was being measured, and the measurements were taken in random order.[12]

(a) Make a table of means and standard deviations for each of the 5×3 combinations of the two factors.

TABLE 15.1 Tool diameter data

Tool	Time	Diameter		
1	1	25.030	25.030	25.032
1	2	25.028	25.028	25.028
1	3	25.026	25.026	25.026
2	1	25.016	25.018	25.016
2	2	25.022	25.020	25.018
2	3	25.016	25.016	25.016
3	1	25.005	25.008	25.006
3	2	25.012	25.012	25.014
3	3	25.010	25.010	25.008
4	1	25.012	25.012	25.012
4	2	25.018	25.020	25.020
4	3	25.010	25.014	25.018
5	1	24.996	24.998	24.998
5	2	25.006	25.006	25.006
5	3	25.000	25.002	24.999

(b) Plot the means and describe how the means vary with tool and time. Note that we expect the tools to have slightly different diameters. These will be adjusted as needed. It is the process of measuring the diameters that is important.

(c) Use a two-way ANOVA to analyze these data. Report the test statistics, degrees of freedom, and *P*-values for the significance tests.

(d) Write a short report summarizing your results.

15.34 Convert from millimeters to inches. Refer to the previous exercise. Multiply each measurement by 0.04 to convert from millimeters to inches. Redo the plots and rerun the ANOVA using the transformed measurements. Summarize what parts of the analysis have changed and what parts have remained the same.

15.35 Discounts and expected prices. Case 15.1 (page 15-18) describes a study designed to determine how the frequency that a supermarket product is promoted at a discount and the size of the discount affect the price that customers expect to pay for the product. In the exercises that followed, we examined the data for two levels of each factor. Table 15.2 gives the complete set of data.

CASE 15.1

(a) Summarize the means and standard deviations in a table and plot the means. Summarize the main features of the plot.

(b) Analyze the data with a two-way ANOVA. Report the results of this analysis.

(c) Using your plot and the ANOVA results, prepare a short report explaining how the expected price depends on the number of promotions and the percent of the discount.

TABLE 15.2 Expected price data

Number of promotions	Percent discount	Expected price									
1	40	4.10	4.50	4.47	4.42	4.56	4.69	4.42	4.17	4.31	4.59
1	30	3.57	3.77	3.90	4.49	4.00	4.66	4.48	4.64	4.31	4.43
1	20	4.94	4.59	4.58	4.48	4.55	4.53	4.59	4.66	4.73	5.24
1	10	5.19	4.88	4.78	4.89	4.69	4.96	5.00	4.93	5.10	4.78
3	40	4.07	4.13	4.25	4.23	4.57	4.33	4.17	4.47	4.60	4.02
3	30	4.20	3.94	4.20	3.88	4.35	3.99	4.01	4.22	3.70	4.48
3	20	4.88	4.80	4.46	4.73	3.96	4.42	4.30	4.68	4.45	4.56
3	10	4.90	5.15	4.68	4.98	4.66	4.46	4.70	4.37	4.69	4.97
5	40	3.89	4.18	3.82	4.09	3.94	4.41	4.14	4.15	4.06	3.90
5	30	3.90	3.77	3.86	4.10	4.10	3.81	3.97	3.67	4.05	3.67
5	20	4.11	4.35	4.17	4.11	4.02	4.41	4.48	3.76	4.66	4.44
5	10	4.31	4.36	4.75	4.62	3.74	4.34	4.52	4.37	4.40	4.52
7	40	3.56	3.91	4.05	3.91	4.11	3.61	3.72	3.69	3.79	3.45
7	30	3.45	4.06	3.35	3.67	3.74	3.80	3.90	4.08	3.52	4.03
7	20	3.89	4.45	3.80	4.15	4.41	3.75	3.98	4.07	4.21	4.23
7	10	4.04	4.22	4.39	3.89	4.26	4.41	4.39	4.52	3.87	4.70

15.36 **Rerun the data as a one-way ANOVA.** Refer to the previous exercise. Rerun the analysis as a one-way ANOVA with $4 \times 4 = 16$ treatments. Summarize the results of this analysis. Use the Bonferroni multiple comparisons procedure to describe combinations of number of promotions and percent discounts that are similar or different.

CASE 15.1

15.37 **ANOVA for chromium and insulin.** Return to the chromium-insulin experiment described in Exercise 15.23. Here is part of the ANOVA table for these data:

Source	Degrees of freedom	Sum of squares	Mean square	F
A (Chromium)		0.00121		
B (Eat)		5.79121		
AB		0.17161		
Error		1.08084		
Total				

(a) In all, 40 rats were used in this experiment. Fill in the missing values in the ANOVA table.

(b) What is the value of the F statistic to test the null hypothesis that there is no interaction? What is its distribution when the null hypothesis is true? Using Table E, find an approximate P-value for this test.

(c) Answer the questions in part (b) for the main effect of Chromium and the main effect of Eat.

(d) What is s_p^2, the within-group variance? What is s_p?

(e) Using what you have learned in this exercise and your answers to Exercise 15.23, summarize the results of this experiment.

15.38 **ANOVA for social insight.** Return to the Chapin Social Insight Test study described in Exercise 15.24. Part of the ANOVA table for these data is given below:

Source	Degrees of freedom	Sum of squares	Mean square	F
A (Gender)		62.40		
B (Group)		1,599.03		
AB				
Error		13,633.29		
Total		15,458.52		

(a) There were 150 individuals tested in each of the groups. Fill in the missing values in the ANOVA table.

(b) What is the value of the F statistic to test the null hypothesis that there is no interaction? What is its distribution when the null hypothesis is true? Using Table E, find an approximate P-value for this test.

(c) Answer the questions in part (b) for the main effect of Gender and the main effect of Group.

(d) What is s_p^2, the within-group variance? What is s_p?

(e) Using what you have learned in this exercise and your answers to Exercise 15.24, summarize the results of this study.

15.39 Do lefties die younger? Do left-handed people live shorter lives than right-handed people? A study of this question examined a sample of 949 death records and contacted next of kin to determine handedness.[13] Note that there are many possible definitions of "left-handed." The researchers examined the effects of different definitions on the results of their analysis and found that their conclusions were not sensitive to the exact definition used. For the results presented here, people were defined to be right-handed if they wrote, drew, and threw a ball with the right hand. All others were defined to be left-handed. People were classified by gender (female or male) and handedness (left or right), and a 2×2 ANOVA was run with the age at death as the response variable. The F statistics were 22.36 (handedness), 37.44 (gender), and 2.10 (interaction). The following marginal mean ages at death (in years) were reported: 77.39 (females), 71.32 (males), 75.00 (right-handed), and 66.03 (left-handed).

(a) For each of the F statistics given above find the degrees of freedom and an approximate P-value. Summarize the results of these tests.

(b) Using the information given, write a short summary of the results of the study.

15.40 Radon detectors. Scientists believe that exposure to the radioactive gas radon is associated with some types of cancers in the respiratory system. Radon from natural sources is present in many homes in the United States. A group of researchers decided to study the problem in dogs because dogs get similar types of cancers and are exposed to environments similar to those of their owners. Radon detectors are available for home monitoring but the researchers wanted to obtain actual measures of the exposure of a sample of dogs. To do this they placed the detectors in holders and attached them to the collars of the dogs. One problem was that the holders might in some way affect the radon readings. The researchers therefore devised a laboratory experiment to study the effects of the holders. Detectors from four series of production were available, so they used a two-way ANOVA design (series with 4 levels and holder with 2, representing the presence or absence of a holder). All detectors were exposed to the same radon source and the radon measure in picocuries per liter was recorded.[14] The F statistic for the effect of series is 7.02, for holder it is 1.96, and for the interaction it is 1.24.

(a) Using Table E or statistical software find approximate P-values for the three test statistics. Summarize the results of these three significance tests.

(b) The mean radon readings for the four series were 330, 303, 302, and 295. The results of the significance test for series were of great concern to the researchers. Explain why.

15.41 Runners versus controls on a treadmill. A study of cardiovascular risk factors compared runners who averaged at least 15 miles per week with a control group described as "generally sedentary." Both men and women were included in the study.[15] The design is a 2×2 ANOVA with the factors group and gender. There were 200 subjects in each of the four combinations. One of the variables measured was the heart rate after 6 minutes of exercise on a treadmill. The observations are in the RUNNERS data set in the Data Appendix. Analyze the data using a two-way ANOVA. Summarize your findings in a short report.

Chapter 15 Case Study Exercises

CASE STUDY 15.1: **Change of majors.** Refer to the data given for the change-of-majors study in the data set MAJORS described in the Data Appendix. Analyze the data for four response variables: SAT verbal score and the three high school grade variables. Each analysis should include a table of sample sizes, means, and standard deviations; Normal quantile plots; a plot of the means; and a two-way ANOVA using sex and major as the factors. Write a short summary of your conclusions.

CASE STUDY 15.2: **Is it economically feasible to grow these plants in a dry climate?** The PLANTS1 data set in the Data Appendix gives the percent of nitrogen in four different species of plants grown in a laboratory. The species are *Leucaena leucocephala, Acacia saligna, Prosopis juliflora,* and *Eucalyptus citriodora*. The researchers who collected these data were interested in commercially growing these plants in parts of the country of Jordan where there is very little rainfall. To examine the effect of water, they varied the amount per day from 50 millimeters (mm) to 650 mm in 100 mm increments. There were 9 plants per species-by-water combination. Because the plants are to be used primarily for animal food, with some parts that can be consumed by people, a high nitrogen content is very desirable. Analyze the data using the methods you learned in this chapter and write a report summarizing your work.

CASE STUDY 15.3: **Additional data on growing plants in dry conditions.** The PLANTS2 data set described in the Data Appendix contains two response variables that were collected according to a two-way ANOVA design. Analyze these data and summarize your work in a report.

Notes for Chapter 15

1. This experiment was conducted by Connie Weaver, Department of Foods and Nutrition, Purdue University.
2. We present the two-way ANOVA model and analysis for the general case in which the sample sizes may be unequal. If the sample sizes vary a great deal, serious complications can arise. There is no longer a single standard ANOVA analysis. Most computer packages offer several options for the computation of the ANOVA table when group counts are unequal. When the counts are approximately equal, all methods give essentially the same results.
3. Bureau of Labor Statistics, *Money Income in the United States, 2000*, Publication P60-213, 2001.
4. This example is based on a study described in P. Payne, "Nutrition adaptation in man: social adjustments and their nutritional implications," in K. Blaxter and J. C. Waterlow (eds.), *Nutrition Adaptation in Man,* Libbey, 1985.
5. Example 15.7 is based on a study described in S. S. Zentall and J. H. Shaw, "Effects of classroom noise on performance and activity of second-grade hyperactive and control children," *Journal of Educational Psychology,* 72 (1980), pp. 830–840.

6. Based on M. U. Kalwani and C. K. Yim, "Consumer price and promotion expectations: an experimental study," *Journal of Marketing Research,* 29 (1992), pp. 90–100.

7. Data from W. M. Smith et al., "The multiple risk factor intervention trial," in H. M. Perry, Jr., and W. M. Smith (eds.), *Mild Hypertension: To Treat or Not to Treat,* New York Academy of Sciences, 1978, pp. 293–308.

8. Data provided by Julie Hendricks and V. J. K. Liu of the Department of Foods and Nutrition, Purdue University.

9. This exercise is based on results reported in H. G. Gough, *The Chapin Social Insight Test,* Consulting Psychologists Press, 1968.

10. See S. Badylak et al., "Marrow-derived cells populate scaffolds composed of xenogeneic extracellular matrix," *Experimental Hematology*, 29 (2001), pp. 1310–1318.

11. Based on A. A. Adish et al., "Effect of consumption of food cooked in iron pots on iron status and growth of young children: a randomised trial," *The Lancet,* 353 (1999), pp. 712–716.

12. Based on a problem from Renée A. Jones and Regina P. Becker, Department of Statistics, Purdue University.

13. For a summary of this study and other research in this area, see Stanley Coren and Diane F. Halpern, "Left-handedness: a marker for decreased survival fitness," *Psychological Bulletin,* 109 (1991), pp. 90–106.

14. Data provided by Neil Zimmerman of the Purdue University School of Health Sciences.

15. Exercise 15.41 is based on a study described in P. D. Wood et al., "Plasma lipoprotein distributions in male and female runners," in P. Milvey (ed.), *The Marathon: Physiological, Medical, Epidemiological, and Psychological Studies,* New York Academy of Sciences, 1977.

Chapter 15

15.1 The response is evaluation of the program's effectiveness. The first factor is the type of training program at $I = 3$ levels. The second factor is how the training is given at $J = 2$ levels. There are $N = 120$ observations.

15.3 The response variable is the comparison of the new lotion with the regular product. The first factor is type of formulation at $I = 5$ levels. The second factor is fragrance at $J = 3$ levels. There are $N = 120$ observations.

15.5 The plot shows that the difference in salaries between men and women does not appear to be the same for each age group. The difference is smallest for the youngest age group, and the gap widens for the next two age groups and then narrows slightly for the age group 65+. This suggests that the two factors interact.

15.7 There is interaction in all plots with the exception of (**c**), for which the two lines are parallel. In (**a**), the response increases more rapidly as the level of Factor A "increases" when $B = 2$ than when $B = 1$. In (**b**), the response increases as the level of A "increases" when $B = 1$ and decreases when $B = 2$. This is the strongest interaction in the four plots. In (**d**), there is no difference in mean response when $B = 1$ and $B = 2$ for the first two levels of A, but there is a large difference at the third level of A.

15.9 The F statistic used to test the null hypothesis of no main effect of packaging has an $F(3, 468)$ distribution. The F statistic used to test the null hypothesis of no main effect of color has an $F(2, 468)$ distribution. The F statistic used to test the null hypothesis of no interaction has an $F(6, 468)$ distribution.

15.11 With 5 promotions, the means are less than with 1 promotion for each level of discount. A discount of 10% gives a higher expected price than a discount of 30% for each number of promotions. The first group, P1-D30 seems more spread out than the others, and there is a low outlier in group P5-D10.

15.13 The means and standard deviations for the treatment combinations are

Variable	Mean	St. dev.
P1-D30	4.2250	0.3860
P1-D10	4.9200	0.1520
P5-D30	3.8900	0.1629
P5-D10	4.3930	0.2685

The means and marginal means are

	% discount		
Number of promotions	10	30	Mean
1	4.9200	4.2250	4.5725
5	4.3930	3.8900	4.1415
Mean	4.6565	4.0575	4.3570

Groups 1 and 4 have much larger standard deviations, with the larger standard deviation in group 4 due to an outlier. The standard deviations are far enough apart that pooling to get MSE is questionable. The tables of means and marginal means suggest that there is little interaction between promotion and discount, but that there are main effects of both factors.

15.15 The information in the basic ANOVA table is contained in the output of all four packages, although the Excel and Minitab output presented are the most basic and simplest to read. To find the sums of squares and tests for main effects and interactions in SAS requires going to the second table in the output. The SPSS output includes several lines at the beginning of the ANOVA table that should not be familiar to you and are of limited use. Excel and SPSS include information on the means and standard deviations for the treatments and the marginal means as part of the basic output, but these can be gotten easily in both SAS and Minitab as well. For just the basic ANOVA table, Minitab produces output most similar to what you have seen in the text.

15.17 **(a)** The response variable is the number of hours of sleep. The first factor is smoking level at $I = 3$ levels. The second factor is gender at $J = 2$ levels. There are $N = 420$ observations. **(b)** The response variable is the measure of concrete strength. The first factor is the formula for the mixture at $I = 6$ levels. The second factor is the number of cycles of freezing and thawing at $J = 3$ levels. There are $N = 162$ observations. **(c)** The response variable is the score on the final exam. The first factor is the teaching method at $I = 4$ levels. The second factor is major at $J = 2$ levels. There are $N = 32$ observations.

15.19 **(a)** The test for the first main effect has an $F(2, 60)$ distribution. The test for the second main effect has an $F(3, 60)$ distribution. The test for interaction has an $F(6, 60)$ distribution. **(b)** At the 5% level, using software, the three F critical values are 3.1504, 2.5781, and 2.2541 for the two main effects and interaction, respectively. **(c)** At the 1% level, using software, the three F critical values are 4.9774, 4.1259, and 3.1187 for the two main effects and interaction, respectively.

15.21 **(a)** You can draw the plot using software or by hand. **(b)** The means for the nonwhites are all slightly higher than for whites. The mean systolic blood pressure increases with age group and there doesn't seem to be an interaction. **(c)** The marginal means are 135.98 for whites and 137.82 for nonwhites. For the age groups the marginal means are 131.65 (35 to 39), 133.25 (40 to 44), 136.20 (45 to 49), 140.35 (50 to 54), and 143.05 (55 to 59). The mean systolic blood pressure is around 2 points higher for nonwhites than whites in each age group. The mean systolic blood pressure increases between 2 and 4 points for each increase in age group.

15.23 (**a**) You can draw the plot using software or by hand. (**b**) The plot suggests that there may be an interaction. The effect of Chromium when going from Low to Normal levels is to decrease mean GITH when the rats could eat as much as they wanted (M) and to increase mean GITH when the total amount the rats could eat was restricted (R). The effect of Chromium appears to be smaller than the effect of Eat. (**c**)

	Eat		Mean
Chromium	M	R	
L	4.545	5.175	4.860
N	4.425	5.317	4.871
Mean	4.485	5.246	4.866

At the Low level of Chromium the difference between M and R is −0.63, and at the Normal level of Chromium the difference between M and R is −0.892. This is reflected in the fact that the two lines are not parallel in the plot.

15.25 The category "other" has the lowest mean SATM score for both males and females. Males in CS have a slightly higher mean SATM score than EO majors, while the females in EO have a higher mean SATM score than the CS majors. From the plot, there appears to be an interaction.

15.27 (**a**) The table gives the sample sizes, means, and standard deviations for the 12 material-time groups. The standard deviations vary considerably, but since each standard deviation is based on only 3 observations, we would expect a large amount of variability.

Variable	Group	N	Mean	St. dev.
% Gpi	ECM1,4	3	65.00	8.66
	ECM1,8	3	63.33	2.89
	ECM2,4	3	63.33	2.89
	ECM2,8	3	63.33	5.77
	ECM3,4	3	73.33	2.89
	ECM3,8	3	73.33	5.77
	MAT1,4	3	23.33	2.89
	MAT1,8	3	21.67	5.77
	MAT2,4	3	6.67	2.89
	MAT2,8	3	6.67	2.89
	MAT3,4	3	11.67	2.89
	MAT3,8	3	10.00	5.00

The following table gives the means for the different material-time groups and the marginal means for material and time.

	ECM1	ECM2	ECM3	MAT1	MAT2	MAT3	Mean
4	65.000	63.333	73.333	23.333	6.667	11.667	40.556
8	63.333	63.333	73.333	21.667	6.667	10.000	39.722
Mean	64.167	63.333	73.333	22.500	6.667	10.833	40.139

(**b**) The most striking feature in the plot is the complete lack of a time effect. The mean % Gpi at 4 and 8 weeks are almost identical, and as a result there is almost no interaction. The important effect is the material, with the ECM (extracellular) material having a much higher % Gpi than the MAT (inert) material. (**c**) The ANOVA table shows no evidence of a time effect ($F = 0.29$, df = (1, 24), $P = 0.595$) or an interaction between time and material ($F = 0.06$, df = (5, 24), $P = 0.998$). However, there is a highly significant effect of material ($F = 251.26$, df = (5, 24), $P < 0.001$).

15.29 The F statistics for the one-way ANOVAs are 58.14 at two weeks, 137.94 at four weeks, and 115.54 at eight weeks. All are highly significant, with P-values less than 0.001. This suggests that there are differences between the materials at each time period. Multiple comparisons can be used to help determine which materials are better at each time period. The general finding from doing the Bonferroni multiple comparisons procedure on all pairs of treatments is that the ECMs are superior to the inert materials.

15.31 Despite the fact that there is a highly significant interaction, for each type of food the mean iron content is higher when that food is cooked in an iron pot than in either clay or aluminum. The main effect of pot is highly significant, and it is clear from the plot of the means that there is little difference between clay and aluminum, while iron pots are associated with higher iron content. The interaction says only that the difference is not necessarily the same between type of pot for each food.

15.33 (**a**) The table gives the means and standard deviations for each of the 15 treatment combinations.

Tool	Time	N	Mean	St. dev.
1	1	3	25.031	0.001
1	2	3	25.028	0.000
1	3	3	25.026	0.000
2	1	3	25.017	0.001
2	2	3	25.020	0.001
2	3	3	25.016	0.000
3	1	3	25.006	0.001
3	2	3	25.013	0.001
3	3	3	25.009	0.001
4	1	3	25.012	0.000
4	2	3	25.019	0.001
4	3	3	25.014	0.002
5	1	3	24.997	0.001
5	2	3	25.006	0.000
5	3	3	25.000	0.001

(**b**) The plot suggests little interaction, as the lines for each time appear fairly parallel. The biggest difference is between the tools, but this is of little importance because we expect the tools to have slightly different diameters and this can be adjusted. There is also a difference in times. With the exception of tool 1, time 2 has the highest diameters, time 3 has the middle diameters, and time 1 has the lowest diameters. (**c**) The df for tool are (4, 30), $F = 412.98$, and $P < 0.001$; the df for time are (2, 30), $F = 43.61$, and $P < 0.001$; the df for interaction are (8, 30), $F = 7.65$, and $P < 0.001$. (**d**) Both main effects and the interaction are statistically significant, although the F for interaction is much smaller than the F for the main effect of time, which is much smaller than the F for the main effect of tool. The interaction occurs because tool 1's mean diameters changed differently over time compared to the other tools. The reason that the interaction is statistically significant despite the appearance of the plot is that the small variability within each treatment makes it possible to detect effects that, although statistically significant, are of little practical importance. Since tool differences are of limited interest, the most important finding is the fact that the diameters vary with time.

15.35 (**a**) The plot shows little interaction, with price generally decreasing with increasing number of promotions and increasing percent discount.

Number of promotions	Percent discount	N	Mean price	St. dev.
1	10	10	4.9200	0.0481
1	20	10	4.6890	0.0737
1	30	10	4.2250	0.1220
1	40	10	4.4230	0.0584
3	10	10	4.7560	0.0768
3	20	10	4.5240	0.0856
3	30	10	4.0970	0.0742
3	40	10	4.2840	0.0645
5	10	10	4.3930	0.0849
5	20	10	4.2510	0.0838
5	30	10	3.8900	0.0515
5	40	10	4.0580	0.0557
7	10	10	4.2690	0.0854
7	20	10	4.0940	0.0761
7	30	10	3.7600	0.0828
7	40	10	3.7800	0.0678

(**b**) The df for promotions are (3, 144), $F = 47.73$, and $P < 0.001$; the df for discount are (3, 144), $F = 47.42$, and $P < 0.001$; the df for interaction are (9, 144), $F = 0.44$, and $P < 0.912$. (**c**) As the number of promotions goes up the expected price tends to decrease. As the percent discount increases the expected price tends to decrease although the sample means of the expected prices for a discount of 40% were above those for a discount of 30% for each of the values of promotion. There appears to be little interaction between the two factors.

15.37 (a)

Source	Degrees of freedom	Sum of squares	Mean square	F
A (Chromium)	1	0.00121	0.00121	0.04
B (Eat)	1	5.79121	5.79121	192.89
AB	1	0.17161	0.17161	5.72
Error	36	1.08084	0.03002	
Total	39	7.04487		

(**b**) The F statistic for interaction has an $F(1, 36)$ distribution and $F = 5.72$. Referring this to the critical values in Table E, we see that the P-value is between 0.025 and 0.010. (Since 36 degrees of freedom is not in the table, we look at the entries for 30 and 40.) (**c**) The main effect of Chromium has an $F(1, 36)$ distribution and $F = 0.04$. From Table E, the P-value is greater than 0.10. The main effect of Eat has an $F(1, 36)$ distribution and $F = 192.89$. From Table E, the P-value is less than 0.001. (**d**) $s_p^2 = 0.03002$, and $s_p = \sqrt{0.03002} = 0.173$. (**e**) The interaction between Eat and Chromium is statistically significant, but the effect is relatively small compared to the main effect of Eat. It would be up to the experimenter to determine whether the difference in the effect of going from Low to Normal levels of Chromium for the two levels of Eat is large enough to be of practical interest. By far, the biggest effect is the main effect of Eat. The mean GITH scores are much smaller when the rats could eat as much as they wanted (M) than when the total amount the rats could eat was restricted (R).

15.39 (**a**) All three of the F statistics have degrees of freedom of 1 and 945. The P-value for interaction is 0.1477, and the P-values for the main effects of handedness and gender were both less than 0.0001. We conclude that there is no significant interaction and that both handedness and gender have an effect on mean lifetime. (**b**) We conclude from the marginal means that women live on the average about six years longer than men, while right-handed people live on the average about nine years longer than left-handed people. No interaction means that the increase in life expectancy for right-handedness is the same for both men and women.

15.41 A plot of the means shows that males have lower mean heart rates than females in both the control and runner groups (by about 15 beats), while the runners have lower heart rates than the controls for both genders (by about 30 beats). Both main effects are highly significant, and the differences reported in parentheses are the differences in the marginal means. A plot suggests little interaction, but the ANOVA table shows that the interaction is statistically significant. Comparing the F-value for interaction with the F-values for main effects, we see that it is much smaller. The reason for the statistical significance of the interaction is that the sample sizes are very large, and with large samples statistical procedures have the ability to detect very small differences. In this case the mean decrease in heart rate associated with running for females is estimated as $148 - 115.99 = 32.01$, while the estimated mean decrease for men is $130 - 103.97 = 26.03$. The analysis tells us that the difference in these decreases is statistically significant, since this is the meaning of interaction. Whether this difference is of practical importance is another issue. If the interaction is judged unimportant, then just summarize these data in terms of the marginal means.